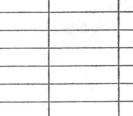

Shocking Science

5,000 years of mishaps and misunderstandings

ILLUSTRATED BY
John Kelly

WRITTEN BY
Steve Parker

Turner Publishing, Inc.

ATLANTA

Library of Congress Cataloging-in-Publication Data
Parker, Steve.
    Shocking science: 5,000 years of mishaps and misunderstandings/illustrated by John Kelly: written by Steve Parker. —1st ed.
        p.   cm.
    ISBN 1-57036-269-6 (a1k. paper)
    1. Science—Anecdotes.   2. Science—History.   I. Title.
Q167.P57   1996
500—dc20   95–18380
                                                                    CIP

Published in the U.S.A. by
Turner Publishing, Inc.
A Subsidiary of Turner Broadcasting System, Inc.
1050 Techwood Drive, N.W.
Atlanta, Georgia 30318

Distributed by Andrews and McMeel
A Universal Press Syndicate Company
4900 Main Street
Kansas City, Missouri 64112

First published in Great Britain 1996
by Hamlyn Children's Books,
an imprint of Reed Children's Books.

First Edition
10 9 8 7 6 5 4 3 2 1

Editor: Andrew Farrow
Designer: Cathy Tincknell
Production Controller: Christine Campbell
Supervising Editor, Turner Publishing: Crawford Barnett
Copy Editor, Turner Publishing: Lauren Emerson
Editorial Assistant, Turner Publishing: Michon Wise

Printed and bound in China
Produced by Mandarin Offset Ltd.

Contents

4
SHOCKING SCIENCE

6
IN THE BEGINNING

8
HEAVENLY MOTION

10
MAPPING THE WORLD

12
DANGEROUS LANDS

14
MONSTROUS PEOPLE

16
MORE MAPPING MADNESS

18
IS THERE LIFE ON MARS?

20
WE ARE NOT ALONE

22
SPACE—THE FINAL FRONTIER

24
IMPOSSIBLE FOSSILS

26
DINO-WARS

28
TRUE OR FALSE?

30
CANE TOADS AND KILLER BEES

32
PESKY PLANTS

34
A HOLE IN THE HEAD

36
TAKING YOUR MEDICINE

38
ALCHEMY

40
TRULY SHOCKING SCIENCE

42
SUFFERING FOR SCIENCE

44
BEYOND THE CALL OF DUTY

46
FLAPPING AND FALLING

48
CHOCKS AWAY!

50
MIS-CONSTRUCTION

52
FALSE STARTS

54
A LIFE ON THE OCEAN—GLUG!

56
INVENTIONS

58
JUST AN ACCIDENT

60
MAD, BAD, AND SAD

62
FAME AT LAST?

64
INDEX

Shocking Science

Scientists are incredible discoverers. They have produced all manner of wonders, from space probes to miraculous medical drugs. Scientists are staid, sensible, solid people. They work methodically and rationally, in search of ultimate truth and understanding. They would never do anything stupid or surprising that would startle or shock us. Oh yeah? Well read on . . .

UNBELIEVABLE

Scientists mess up occasionally, like everyone else. They have misunderstood the world about them, ignored evidence, lusted after fame and fortune, and just plain gotten it wrong. This book takes the lid off science through the ages, to reveal blunders, hoaxes, unlucky accidents, and bizarre beliefs. Some were harmless, and now are laughable. Others were truly shocking and resulted in wasted time and effort, and even tragic loss of life.

UNDERSTANDING

The first people who tried to understand the universe were called "natural philosophers." They studied stars, rocks, plants, animals, and other aspects of the natural world. Modern branches of science, like astronomy, physics, and genetics, developed because these scholars tried to find out about the world they lived in.

Now we can look back and smile at the odd ideas and funny beliefs of the past. In years to come, will people smile at us, with our odd ideas and strange beliefs?

It's just a program about the Stone Age

4

UNFORGETTABLE?

Even a towering scientific genius like Isaac Newton was at the mercy of fate's little twists. He kept careful notes and mathematical equations—but it is said that his dog knocked over a candle and burned the papers. Isaac therefore had to rely on memory when writing his monumental book on mathematics, *Principia*, which revolutionized much of science. If his dog had been less clumsy, might science have progressed even further?

UNFORGIVABLE

In the 1970s, scientists were startled to hear that a Russian researcher had successfully transplanted furry skin from black mice to white mice. What an amazing breakthrough for medical science! If a living body could be prevented from rejecting new tissue, transplants would be more successful. Then it was discovered the researcher was a fraud. He had drawn black patches on the white mice with a felt-tip pen.

UNREPEATABLE

In the early 1990s, two scientists reported that they had succeeded with "cold fusion." They could produce energy from a test tube of special water, without the drawbacks of the usual "hot fission" used in nuclear power stations. Limitless power from a cup of cold water? An end to all our energy and pollution problems? Despite hours of experiments and millions of dollars, nobody else has managed to repeat the results.

I think there's something you should know

In the Beginning

I n the beginning, what was there? A black hole? A supreme being? A giant dragon? Through the ages, legends and religions have tried to explain the beginning of the universe and the creation of our own Earth and Sun. Today's scientists think they are much nearer to the correct answer. But hundreds of years ago, scientists thought *they* were near the truth—and their ideas were very different . . .

The Cosmic Egg

In ancient Greece, some people believed that the gods created the heavens and Earth. Others said the world had always existed. Nearly 2,000 years ago, a Chinese myth told how the world was sculpted by a newly hatched giant. "First there was the great cosmic egg. Inside . . . was P'an Ku, the Divine Embryo. And P'an Ku burst from the egg. . . . With a hammer and chisel in his hands, he fashioned the world." The idea that the Earth hatched or grew from an egg is common to many ancient beliefs.

30 days hath September...

A Nine o'clock Start

The Bible tells how the heavens and earth and all living things were created by God in six days. In 1650, Irish bishop James Ussher examined the listings of people in the Bible and calculated their dates of birth. His work was refined by John Lightfoot into an official timetable of Creation. It claimed the Earth was made at nine o'clock in the morning on October 26, in 4004 B.C. This information was added to some Bibles, and it was accepted as fact by a great number of people for many years.

SUDDENLY GROWING OLD

Until the early 1800s, most scientists believed in the Catastrophe Theory. This stated that the Earth and its oceans and mountains had been formed by great floods and other catastrophes described in the Bible. But Scottish geologist James Hutton had another theory that he derived after studying how slowly rocks formed and wore away. He guessed that because this had been happening since the Earth began, the Earth must be *incredibly* old—this was contrary to many religious beliefs of the time. Hutton's *Theory of the Earth* was published in 1795, but the book was so boring, and its suggestions seemed so crazy, that most people ignored it.

OLDER AND OLDER

A few scientists did expand upon Hutton's ideas, and by the 1850s many believed the Earth was more than 6,000 years old.

⊕ In the mid-1700s, French naturalist Comte de Buffon said the Earth had cooled from red-hot rocks. By testing the cooling rate of iron balls, he estimated it was 75,000 years old.

🎓 Later, in the mid-1800s, Scottish physicist William Thomson calculated the Earth's age as nearer to 100 million years.

⚛ By 1907 the estimated age of the Earth had leaped to 410 million years. And in the 1930s it was thought to be at least 1,000 million years old.

1 0 0 0 0 0 0 0 0

THE BIGGEST BANG

Today, scientists think that the universe came into being about 15,000 million years ago. It began when a tiny speck containing all matter blew up in a "Big Bang." The Sun, Earth, and other planets formed around 4,600 million years ago, probably from clouds of gas and dust whirling in space.

At least, that's the *current* view . . .

Heavenly Motion

For thousands of years, people have lain on their backs and gazed up at the night sky. "Are the Moon and stars small and near," they ask, "or large and far away? Why do they move in the same regular pattern? And why do . . . *yawn* . . . *snore* . . ." (These great mysteries are a wonderful cure for sleeplessness.)

PERFECT CIRCLES

From earliest recorded history, most people assumed the Earth was at the center of everything. The Sun, Moon, and stars revolved around it. And why not? It certainly looked that way. Ancient Greeks such as Aristotle believed that stars and other heavenly bodies moved in perfect circles. And they felt there could be no movement unless one thing provided the moving force—God.

CRYSTAL BALLS

In ancient Egypt, the scientist Ptolemy added to Greek ideas on astronomy. He said the Sun, Moon, and stars were fixed to crystal spheres—vast, thin-walled balls of a clear substance. But to fit with observations of the stars' movement, Ptolemy needed eighty crystal spheres, all turning in different directions! Still, lots of people liked the idea of Ptolemy's perfect spheres, thinking that God would surely have made a heavenly system as harmonious as this. This Earth-centered system dominated astronomy for 1,400 years.

The Sun at the Center

In the early 1500s, Polish astronomer Nicolaus Copernicus said Earth (and other planets) went round the Sun! His book, *De revolutionibus orbium coelestium*, published in 1543, even said that Earth spun like a top! This Sun-centered system was much simpler and more sensible than Ptolemy's complicated spheres. But the Roman Catholic Church was horrified. It believed that God had made Earth as the center of the universe. It forbade people to even mention Copernicus's ideas.

Oval Orbits

While studying the planet Mars, German scientist Johannes Kepler realized that Earth and other planets *did* orbit the Sun—only not in perfect circles. In 1609 he proposed that the planets moved around the Sun in ellipses, or ovals.

A Moving Earth

Until this time, nobody had used a telescope, because it wasn't invented until 1608! When the brilliant Italian scientist Galileo Galilei finally gazed through one, he saw mountains on the Moon, stars too faint for the naked eye to see, and four moons revolving around the giant planet Jupiter.

This was the first direct proof that Earth was not the center of everything. So Galileo spoke out in support of Copernicus. In 1633 the Church summoned him to Rome where he was told to admit that Earth *was* the center of the universe and *did not move*. He was forced to agree, although legend says that as he left, he muttered about Earth: "But it *does* move." (Actually, he would have said *Eppure si muove*, since he was Italian.)

Final Apology

Galileo was sentenced to house arrest for life. The Church ordered his books burned, and banned his teachings. Exactly 360 years later the Pope and Catholic Church finally apologized and admitted that Galileo had been right all along.

Mapping the World

In ancient times, when most people rarely traveled beyond their own town or village, a few scholars said that the earth was round. What a crazy idea! Everyone knew the earth was flat . . .

CENTERED ON THE MED

In the time of the Greek poet Homer, almost 3,000 years ago, maps showed the Mediterranean Sea sitting in the center of a flat world. Empty lands surrounded it and were in turn surrounded by ocean. What lay beyond that . . . no one knew!

STICKS AND SHADOWS

Then, about 2,200 years ago in ancient Egypt, the poet and scientist Eratosthenes became curious about a small problem. Each year at noon on June 21, in the city of Syene, the Sun was directly overhead. Its rays shone straight down a deep well. Yet in Alexandria, to the north, the rays did not shine straight down. They cast shadows on a tall obelisk.

Eratosthenes reasoned the earth was not flat, but perhaps was round like a ball. So he hired a man to pace out the distance between the two cities: 800 km. From this, Eratosthenes calculated it was 40,000 km around the earth. He was not far off. But gradually many people forgot his work and went back to the idea of a flat earth.

20,752,
20,753,
20,7... err...
oh,
#~**>#!

A ROUND WORLD

About 300 years later, the Egyptian scientist Ptolemy refined some Greek ideas. In his book *Geography*, he returned to the idea of a ball-shaped earth. Europe, North Africa, and the Middle East now had a fairly accurate shape. But Ptolemy's map did not show any land below the equator, which he drew too far north.

TAKING SHAPE

By the medieval period, about A.D. 1000, Arab geographers had used information from sailors and merchants to create a fairly accurate map of the northern world. They added details of China and other Eastern lands where they traded silks and spices. Although they did not know about the Americas, they suspected there was a great southern continent, a combination of Australia and Antarctica!

RISKY ROUND THE EDGES

Meanwhile, in Europe, maps had taken several steps backwards. They were usually tied to religion and superstition. Most depicted a flat earth with the holy city of Jerusalem at the center. Many maps showed God looking down on the world. In the less well known areas to the east and south, mapmakers drew demons and fierce warriors waiting to invade Christian lands on behalf of the Devil.

How about this?

NEARLY ROUND

By the time Christopher Columbus discovered the Americas in 1492, people had figured out that the earth was round. We now know the earth is not a perfect sphere. The distance round the equator (40,075 km) is 67 km more than the distance around the Poles. And the North Pole is 45 meters taller than the South Pole.

Dangerous Lands

In ancient times, people believed they were living at the center of a flat earth. They thought that beyond their own region lay strange oceans and dangerous lands . . .

SEAS WITH DRAINS

Many early peoples thought that their lands were surrounded by a great river or ocean. Far across it, the sea poured into a vast whirlpool, down a gigantic drain, and into the blackness. The idea of a whirlpool made sense. Otherwise, people said, the rain would fill up the oceans, and they would eventually flood the land.

HERE BE DRAGONS!

Some people believed that distant seas and lands were populated by massive dragons and serpents that breathed fire and ate unlucky travelers for breakfast. Sometimes they saw the dragon's smoke and fire far in the distance, and heard it roaring. Today scientists would say that the rumbling and smoke probably came from a volcano.

WHEELS AND MOTIONS

Medieval storytellers described how a curious man traveled to the end of the flat earth and poked his head through the curtain that was the sky. On the other side he saw the wheels and gears and levers that moved the Sun, Moon, and stars. But the gods were angry at this intrusion. They quickly mended the curtain, trapping the man by the neck, and he died.

How very interesting

THE SUN'S JOURNEY

In ancient Egypt, there were many beliefs about the Sun. One said that it was pushed across the sky by an enormous scarab beetle. It was thought to be swallowed at dusk by the sky goddess, Nut, who passed it through her body during the night and gave birth to it again the next morning.

In a box

Cosmas, an Egyptian geographer, imagined that the world was contained in a huge box, with the heavens in its bulging lid. He believed day and night were created by a huge mountain that obscured the Sun for part of its journey.

Oh well, here we go again...

SONGS FROM THE SEA

In Greek legends, sailors were lured to their deaths by the beautiful singing of three sirens (creatures that had the upper body of a woman and the lower body of a bird). Attracted by the songs, the men would crash their ships in the rocky waters surrounding the sirens' island and drown. In European folklore, some mermaids (who had the upper body of a woman and the lower body and tail of a fish) also sang lovely but deadly songs. Others were thought to be kind, saving sailors from shipwrecks.

More blood and stones anyone?

Monstrous People

In the sixteenth century, a person on the island of Ceylon (now Sri Lanka) saw a strange sight: "In harbor are some very white and beautiful people who wear boots and hats of iron and never stay in any place. They eat a sort of white stone and drink blood." Were they demons from hell? Servants of the gods? Or could they have been Europeans wearing metal helmets, eating bread, and drinking wine? . . .

PLINY'S PEOPLE

One of the most important scholars of Roman times was Pliny the Elder. At dinner parties, Pliny would "entertain" his guests with some of the 20,000 facts in his 37-volume encyclopedia *Historia Naturalis*. These "facts" included tall tales of peoples that lived in distant lands—such as the sciopods, who used their one huge foot to shade themselves from the Sun. There were also people with dogs' heads (called cynocephali); faces on their chests (the blemmyae, who lived in Africa); and even ones with huge ears that they slept in and used for flying (the panotii, of the All Ears Islands).

What is it?

SCHOLARLY DEBATE

The possible existence of these peoples kept scholars arguing for centuries. In the twelfth century, the Church even debated if it was possible to convert the dog-headed people of India to Christianity! The books of epic Venetian traveler Marco Polo, from the late thirteenth century, also included amazing tales of humanoid monsters. But, as always, there was no proof.

A STRANGE NEW WORLD?

In 1492, Christopher Columbus made his first great voyage of exploration. Instead of reaching his original destination, India, he found a "New World" of lands not seen on the maps—the Americas. Columbus found no monsters in this world. When describing the natives whom he met, Columbus wrote: "The people are very well formed, with handsome bodies and very fine faces."

Despite the continuing lack of evidence, people in Europe still wanted to believe in human monsters, and scholars continued to tell stories about hideous creatures living on other continents. Even in the mid-sixteenth century, books continued to show sciopods living in India!

HUGE, HAIRY BRUTES

Even today, according to some Himalayan peoples, the Yeti, or Abominable Snowman, roams their snowy mountains. In 1921, English explorers claimed to have found this huge, hairy creature's footprints. A similar beast known as Sasquatch, or Bigfoot, is supposed to live in the forests of northwest North America and the swamps of southeastern U.S.A. The only evidence, however, includes footprints, some shadowy photographs, and bits of fur.

15

More Mapping Madness

In about A.D. 1000, Egyptian scientist Alhazen suggested to his Caliph (ruler) that they build a dam across the upper River Nile. This would be used to store water for crops. The Caliph agreed to this wonderful idea, and he promptly sent Alhazen on an expedition to make maps and plans. But Alhazen was just one of a great number of scientists who have discovered that many things are easier said than done.

MAD MAPS

Unfortunately for Alhazen, his survey showed that the dam would be impossible to build. In fact, he feared the Caliph's wrath so much that he pretended to be mad—*for nearly twenty years*. Then, upon the Caliph's death, Alhazen made a sudden recovery and went on to become a famous physicist, doing brilliant work on the properties of light.

SCALING THE HEIGHTS

Even with modern mapmaking equipment, scientists still do not always agree:

1860 British scientists measure Mount Everest in the Himalayas. They say it's 8,840 meters high and is the world's tallest peak.

1880 The second highest peak is declared as K2 in northern Pakistan, at 8,611 meters.

1973 Chinese mapmakers revise Mount Everest's height to 8,848 meters.

1987 In **March**, U.S. satellites measure K2 as at least 8,858 meters—it's now said to be taller than Everest! In **August**, the Chinese disagree, saying Everest is still 8,848 meters and K2 is 8,611 meters. In **October**, more satellites measure Everest as being even taller, at 8,863 meters. But K2 gets smaller, at 8,607 meters.

Let's leave the arguments there, and have a look at the theory of "Continental Drift."

DRIFTING APART

The theory of Continental Drift was first suggested by Alfred Wegener in 1912. He said that there was once only one continent on Earth. About 200 million years ago, it split up and the earth's land masses gradually drifted apart. Experts scoffed at this theory in the 1920s. Some insisted Wegener be banned from writing scientific articles or attending scientific meetings. Why such bad feeling?

✄ Wegener was a meteorologist, or weather expert. Geologists claimed to be the only real earth scientists. They did not like the idea of a mere weatherman interfering with their science.

✎ Wegener was German, and Germany had been the enemy of many countries in World War I.

However, from the 1930s more and more evidence supported Wegener's theory, and by the 1970s the theory of Continental Drift had been accepted.

17

Is There Life on Mars?

In the 1870s, Italian astronomer Giovanni Schiaparelli studied the planet Mars through his telescope. He thought that he saw lines across the surface. He called them *canali*, an Italian word meaning channels or canyons. But the word became translated as "canals" in English, giving people the idea that someone had made them.

When American astronomer Percival Lowell observed the *canali*, he also concluded they were too straight to be natural. In 1895 Lowell wrote that intelligent creatures had dug the "canals" to bring water from the planet's frozen poles to the desert-like central regions. More recently, some observers have claimed that a huge, face-shaped area seen in photos is part of a Martian city, complete with avenues and a pyramid!

FLY-BYS AND ORBITS

In 1976, space probes *Viking 1* and *Viking 2* landed on Mars. Their photos failed to show any canals. The Martian landscape consisted of dead volcanoes, red rocks, deep valleys, and violent dust storms. It seems the canals were merely an optical illusion. The results of experiments to detect life were also uncertain. Scientists wanted more proof, however, so in 1992 they launched the multimillion-dollar probe, the *Mars Observer*. . . .

BEEN AND GONE—MISSING MISSIONS

In August 1993, the *Mars Observer* was about to enter orbit around Mars. Its task was to study the planet's surface and atmosphere in great detail. Suddenly its radio signals ceased. To this day, no one is certain what happened to the *Observer*.

The *Mars Observer* was not the first Mars spacecraft to fail in its mission. The Soviet Union (now Russia) also launched many unsuccessful missions to the red planet:

Mars 1 took off in 1962, but its instruments failed during the journey. Two sister craft went off course in the same year and disappeared into space.

In 1964, *Zond 2* stopped sending signals on its journey to Mars.

In 1971, *Mars 2* and *Mars 3* went into orbit and dropped probes to the surface. These went silent on the descent, lost in the biggest dust storm ever seen on Mars.

Four more missions went to Mars in 1973. One went off course. One failed to slow down and flew past the planet and into deep space. One failed on its descent to the surface. Finally, the last Soviet mission sent back a few vague pictures.

IS ANYONE THERE?

In the 1890s, Guglielmo Marconi invented the radio. Suddenly, listening for signals from outer space became the latest craze. In 1901, 100,000 francs were offered for the first person to speak with aliens. Martians did not count, however, because most people were already convinced of their existence.

19

We Are Not Alone

Many scientists agree—we are not alone. There's a good chance that life of some kind exists elsewhere in the universe. Hundreds of people claim to have seen strange spacecraft and creatures from other planets, and millions of dollars have been spent studying UFO sightings. But, so far, there's no proof. . . .

Look—a reply from the gods!

FIRST VISITORS?

History is littered with legends of powerful beings who inhabit the skies. In Peru, the Nazca desert is marked with patterns and lines several kilometers long. Were they messages made by ancient Nazca people to communicate with their gods or visiting aliens? Some people claim that the coffin of a Mayan king, Pacal, shows a spaceman who visited his city of Palenque, in Mexico, 1,300 years ago.

FIRST MODERN SIGHTING?

The modern age of UFOs—Unidentified Flying Objects—began in 1897 in Le Roy, Kansas. A farmer claimed he saw a huge torpedo-shaped craft with a glass cabin below, occupied by six strange beings. It landed, lassoed one of his calves with a rope, and took to the skies, trailing the unlucky beast heavenward!

20

FIRST MEETING?

In his 1953 book *Flying Saucers Have Landed*, American George Adamski said he had met a being from Venus, in the California desert. It was human-shaped with green eyes and long blond hair. Adamski said he later met other aliens and traveled with them to Venus, Mars, Jupiter, and Saturn.

SAUCER PRINTS

For many years experts were puzzled by circles of flattened crops that appeared in fields. Did these corn circles show where "flying saucers" had landed? When two men admitted they had created the circles, some experts refused to believe them!

First UFO fatality

In January of 1948, staff at the Goldman Airfield in Kentucky saw a bright disc-shaped object high above. Captain Thomas Mantell flew his P-51 Mustang fighter to investigate. He flew too high, blacked out from lack of oxygen and crashed to his death. Did he get carried away pursuing a UFO— or did the foreign ship foil his attempts to follow it?

SEEING IS BELIEVING

Many photos of UFOs are hoaxes. Some show car hubcaps being thrown like frisbees, or are reflections in windows. And scientists have many other explanations for UFO sightings:

✦ Secret planes being tested at night.
〰 Reflections in the sky caused by layers of hot air, like desert mirages.
⚕ Satellites and spacecraft reflecting the sun, or debris and meteors burning up in the atmosphere.
✦ Earthquakes or powerful storms causing unusual brainwaves in some people. These brainwaves produce sensations such as floating, flying, leaving the body, and being touched.

Not now! I'm looking for UFOs!

Space–The Final Frontier

Space travel is one of the most recent sciences. In October 1957, a Russian rocket blasted the world's first artificial satellite, *Sputnik 1*, into orbit. A month later, *Sputnik 2* carried the first space traveler, a dog named Laika, and on April 12, 1961, Russian cosmonaut Yuri Gagarin became the first spaceman. Since then, hundreds of people have flown into space, even to the Moon, and returned safely to Earth. But in this most exacting of sciences, there has also been the occasional blunder.

Walkies! Floaties!

TILES FROM HEAVEN?

The NASA Space Shuttle blasts into space with the help of two booster rockets and a huge fuel tank for its engines. After the mission, it glides back into the atmosphere and lands. Special tiles on its underside shield it against the enormous heat caused by the friction of reentry into the atmosphere. That is, unless the tiles fall off. In early tests, the glue did not work properly, and the tiles had been slapped on carelessly by students working summer jobs!

Err... Houston... I think we have a problem.

GET ME ANOTHER WRENCH!

In 1985, shuttle astronauts repairing a faulty satellite discovered that its outer nuts were too big for their wrenches. It turns out that the satellite's makers had changed the design without telling the space agency. It's not easy to dash out to the local hardware store and get another wrench when you're floating 200 km out in space!

22

PUTTING OUT THE TRASH

The Russian orbiting space station *Mir*, launched in 1986, has been visited by the manned *Soyuz* spacecraft, the automated *Progress* craft carrying supplies, and even the Space Shuttle. Spacecraft lock onto *Mir's* docking ports, and cosmonauts can crawl through a linking tunnel. But when a spacewalking cosmonaut first tried to link a special *Kvant* science module to *Mir*, he found that the docking port was blocked by a plastic sack full of trash from the factory!

HUBBLE TROUBLE

In 1990, the Space Shuttle launched the U.S.A.'s orbiting Hubble space telescope. High above the atmosphere, it would give astronomers clear, sharp views of stars in deep space, much better than could be obtained from Earth.

But eager scientists soon found that the pictures were blurred! Hubble's mirror was out of shape—it had been polished beautifully, but to the wrong curvature. Experts had noticed the error in tests, but it was so large, they thought the measuring devices were faulty. So NASA had to send out another Shuttle mission to fix the lens and correct Hubble's sight.

Impossible Fossils

Fossils are the only traces that remain of animals and plants that have been dead for thousands of years. When people discovered increasing numbers of fossils in the nineteenth century, they soon realized that many types of plants and creatures had died out.

It's so nice to be out of that smelly old Ark

DROWNED IN THE FLOOD

The Bible says God created all life on Earth. So it was decided that fossils must be the remains of living things that had been drowned in the Great Flood. The species of animals that survive today were said to be descendants of those saved by Noah and his Ark. But then more fossils were found even deeper in the ground. So the story was changed to two floods, then several, then perhaps seven or more.

One of the first fossil experts, Baron Georges Cuvier of France, said that God had been dissatisfied with the living things on Earth. So He destroyed them with catastrophes such as earthquakes and floods and then created a new batch of plants and animals. This supposedly happened many times.

CORKSCREW HOLES

Gold diggers in the Wild West of the U.S.A. kept finding strange twisty holes in the soil, which they nicknamed "Devil's corkscrews." The burrows went more than two meters into the ground. Scientists were puzzled. They suggested the holes had been made by twisted tree roots, or even blasts of lightning. The real answer is almost as strange. The burrows were made by a pig-sized beaver called the *Paleocastor*, which lived fifteen million years ago.

BEAVER HOMES

HEADS OR TAILS?

Fossil-hunting is a tough business. You kneel on hard rocks in the dust and heat, bashing your fingers with the hammer as you chip at stone. At last—you find a fossil! So you take it home and start to rebuild the original animal. After months of hard work you decide it was a beast with two heads, curly horns, three legs with hooves and two more with claws, and a tail on its chest. So why does everyone laugh?

The Rocky Mountains in Canada contain thousands of animals preserved as fossils, including one named *Anomalocaris*. Scientists once thought that its parts came from many different animals. Its circular mouth was believed to be from a jellyfish, its long claw was said to be from a prawn, and its body was thought to be that of a sea cucumber or sponge. We now think that it was a "killer shrimp" one meter long, with two great spiny feelers.

HALLUCIGENIA

Not all strange reconstructions are mistakes. The extraordinary wormlike *Hallucigenia* was so named because the scientist who first studied it thought he was seeing things. Only a few centimeters long, it had seven pairs of spines and seven soft, bendable tubes along its body. Scientists are still not sure which part of this animal's body represents its head or tail.

Dino-wars

Dinosaurs were prehistoric reptiles. They first appeared about 230 million years ago, and all had died out by 65 million years ago. It was not until the 1820s, when doctor Gideon Mantell and his wife found some fossilized teeth and bones, that people began to study dinosaurs seriously.

MANTELL'S LIZARD

One expert declared that Mantell's finds came from an ancient rhinoceros. Mantell, however, decided that they were from a plant-eating lizard, similar to an iguana but thirty-five meters long. His reconstruction, the first ever made of a dinosaur, showed that it walked on all fours and had a spike on its nose. He called his find an *Iguanodon*, or "iguana tooth." We now know that the *Iguanodon* was ten meters long, probably walked on its hind legs, and its nose-horn was really a thumb bone.

HEADS OR TAILS?

In the 1870s, many dinosaur fossils were found by American scientists Edward Drinker Cope and Othniel Charles Marsh, who were bitter rivals. In 1868 Cope had assembled the fossils of a swimming reptile called a plesiosaur. When Marsh saw the reconstruction, he pointed out that Cope had put the head in the wrong place—at the end of the dinosaur's tail!

THE BONE WARS

The rivalry between Cope and Marsh was so great that they began a race to find dinosaur fossils in the American West—which didn't seem big enough for both of them. In 1878, at Como Bluff, Wyoming, some men digging fossils for Marsh saw two strangers beginning to dig lower down the slope. Thinking the strangers were Cope's men, they deliberately set off a rock slide and forced the strangers to flee!

The two teams found so many fossils—and discovered about two hundred new species—that they couldn't carry them all. They smashed them up, rather than leave them for other expeditions to study.

HOW THE *BRONTOSAURUS* DIED TWICE

In 1877 Cope decided some fossils belonged to a huge new dinosaur which he named *Apatosaurus*. In 1879 Marsh found another set of bones and called them *Brontosaurus*, "thunder lizard." After many years of debate, experts agreed the two sets of bones belonged to the same type of dinosaur. So the dinosaur is now called *Apatosaurus*, after the first bones named. Thus the *Brontosaurus* has died out twice, in life about 140 million years ago, and in name in 1960.

A FINAL INDIGNITY

For many years the *Apatosaurus* was rebuilt with the wrong head. Experts mistakenly used the skull of another dinosaur, called *Camarasaurus*, in reconstructing it. This head was tall, with leaf-shaped teeth. The real skull of *Apatosaurus* was identified in 1975. Its head was low, with peglike teeth.

True or False?

The world of science is occasionally startled by a new and unexpected discovery. Then it is stunned, even shocked, if it finds that the discovery is a hoax or trick. In 1726, when German doctor Johann Beringer wrote a book about several new types of fossils he had discovered, he was understandably excited. Perhaps he would become famous! Unfortunately, the fossils were fakes, planted by rival scientists.

...and here's one I made earlier

THE "MISSING LINK"

In 1912, amateur fossil hunter Charles Dawson found bones and tools in Piltdown, England. There were parts of a human skull, and even a lower jaw with teeth. British experts were overjoyed. The discovery became known as the "Piltdown Man," the "missing link" in the evolution of humans from apes that lived two million years ago. His domed skull, like that of a modern human, housed his big brain. His strong, apelike jaws and teeth were left from his evolutionary past.

In 1953, chemical dating showed that the Piltdown Man was a forgery. The skull bones were human, but only a few hundred years old. The lower jaw and teeth were from an orangutan—filed down, colored to look prehistoric, and placed alongside the bones. Who could have done this? No one knew. But the Piltdown Man had so well matched the expectations of many scientists that for years they failed to check the evidence properly.

BEAKS AND FUR

In 1799, scientists were presented with the dried skin of a strange Australian animal. It had a furry body like a water vole, but a leathery beak and webbed feet like a duck! The experts were sure it was fake, since they had often been shown amazing creatures that turned out to be the sewn-together skins of different animals.

The experts tried to pry off the beak, but it would not budge! This was no fake. It was the first platypus to be examined by scientists. It was a mammal with fur, and it produced milk for its babies. But it also had a birdlike beak, webbed feet, and it laid eggs.

Frankly, I just can't see what's so funny

THE SEARCH FOR NESSIE

For centuries, tales have been told of a large creature living in Loch Ness, Scotland's deepest lake. The first recorded sighting was in A.D. 565. Saint Columba said he'd attended the burial of a man apparently bitten by a huge monster while swimming in the lake.

In the 1930s, sightings of Nessie increased when a new road was built around the shores of the lake. Some people said the monster resembled a plesiosaur, a twelve-meter water reptile from the age of dinosaurs. Others said it looked like a huge snake. Early photos showed dark shapes moving across the lake in the distance. In the 1970s, an under-water photo showed a bulky shape with a diamond-shaped flipper. But Nessie has still not been found. Some photos have been exposed as fakes, others are more likely pictures of waves or floating logs. Recent expeditions using submarines and sonar (echolocation) have found no trace of the monster. But tourists still flock to the lake and even hire helicopters in the hope of spotting it.

29

Cane Toads and Killer Bees

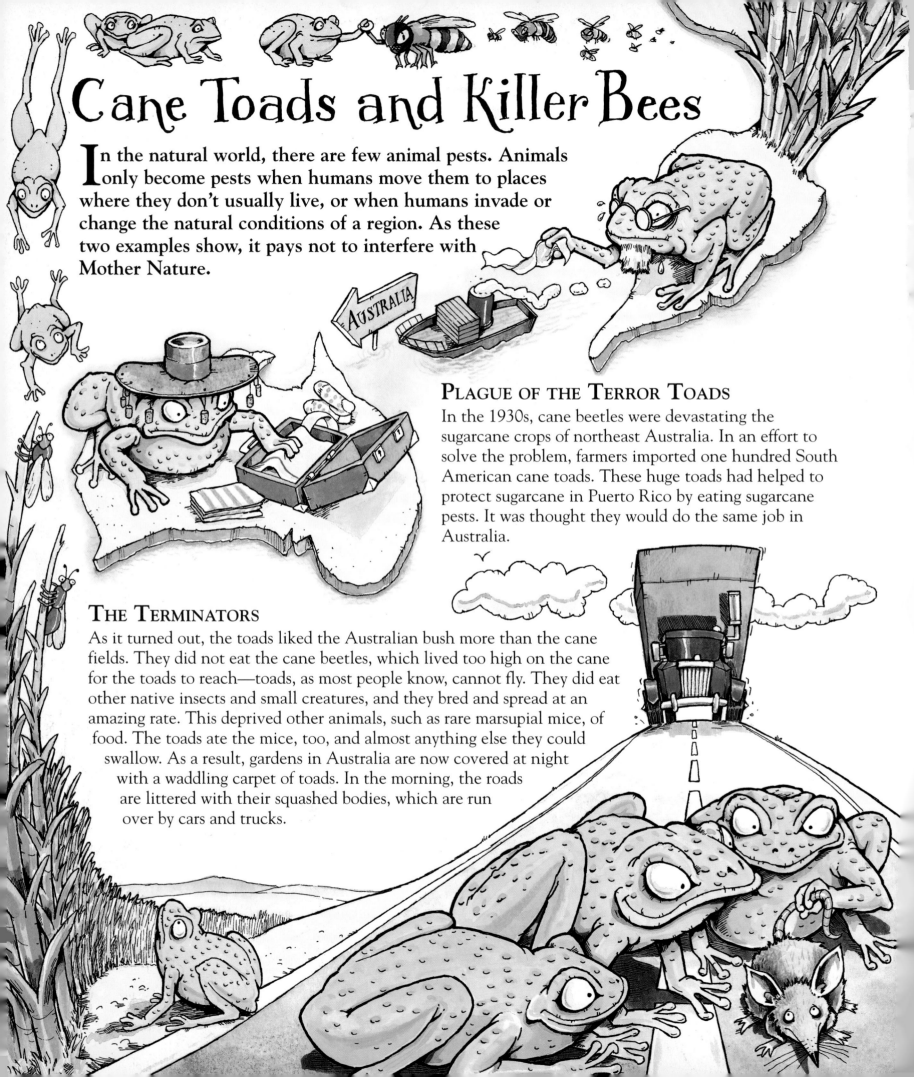

In the natural world, there are few animal pests. Animals only become pests when humans move them to places where they don't usually live, or when humans invade or change the natural conditions of a region. As these two examples show, it pays not to interfere with Mother Nature.

PLAGUE OF THE TERROR TOADS

In the 1930s, cane beetles were devastating the sugarcane crops of northeast Australia. In an effort to solve the problem, farmers imported one hundred South American cane toads. These huge toads had helped to protect sugarcane in Puerto Rico by eating sugarcane pests. It was thought they would do the same job in Australia.

THE TERMINATORS

As it turned out, the toads liked the Australian bush more than the cane fields. They did not eat the cane beetles, which lived too high on the cane for the toads to reach—toads, as most people know, cannot fly. They did eat other native insects and small creatures, and they bred and spread at an amazing rate. This deprived other animals, such as rare marsupial mice, of food. The toads ate the mice, too, and almost anything else they could swallow. As a result, gardens in Australia are now covered at night with a waddling carpet of toads. In the morning, the roads are littered with their squashed bodies, which are run over by cars and trucks.

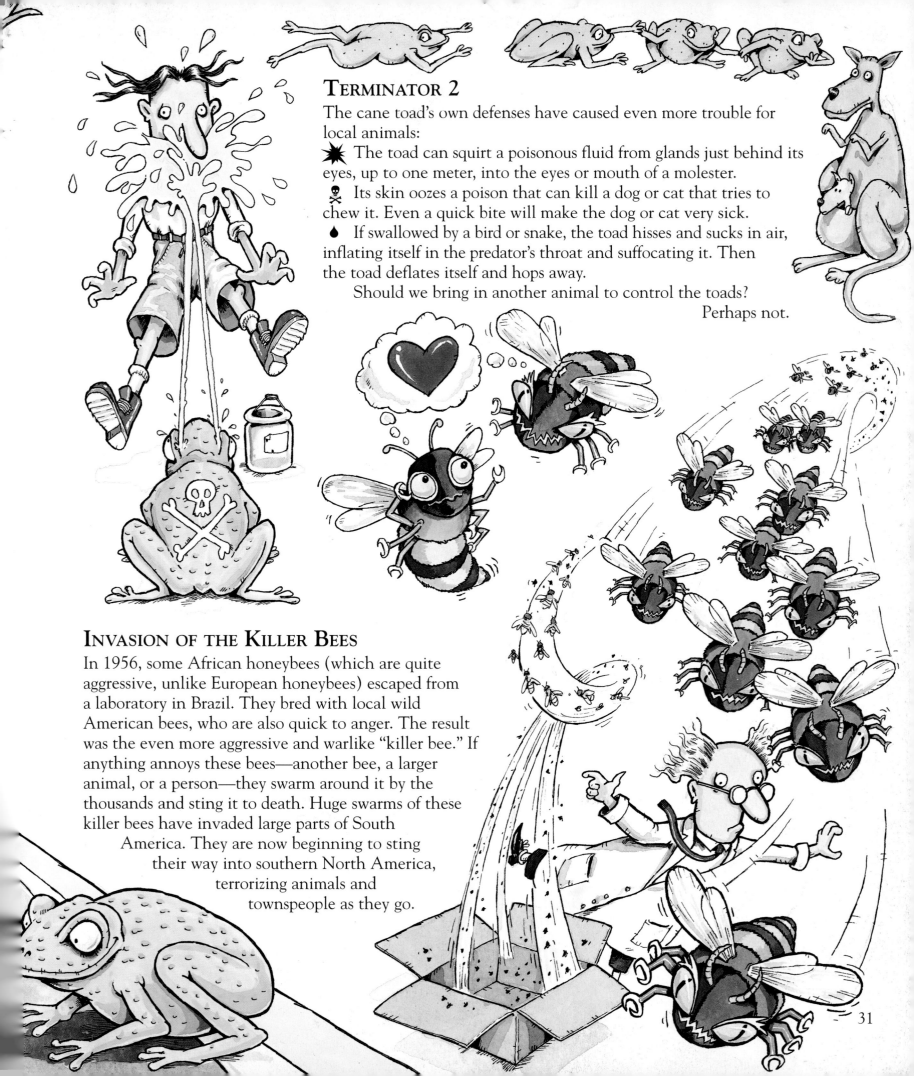

TERMINATOR 2

The cane toad's own defenses have caused even more trouble for local animals:

💥 The toad can squirt a poisonous fluid from glands just behind its eyes, up to one meter, into the eyes or mouth of a molester.

☠ Its skin oozes a poison that can kill a dog or cat that tries to chew it. Even a quick bite will make the dog or cat very sick.

🌢 If swallowed by a bird or snake, the toad hisses and sucks in air, inflating itself in the predator's throat and suffocating it. Then the toad deflates itself and hops away.

Should we bring in another animal to control the toads?

Perhaps not.

INVASION OF THE KILLER BEES

In 1956, some African honeybees (which are quite aggressive, unlike European honeybees) escaped from a laboratory in Brazil. They bred with local wild American bees, who are also quick to anger. The result was the even more aggressive and warlike "killer bee." If anything annoys these bees—another bee, a larger animal, or a person—they swarm around it by the thousands and sting it to death. Huge swarms of these killer bees have invaded large parts of South America. They are now beginning to sting their way into southern North America, terrorizing animals and townspeople as they go.

31

Pesky Plants

Plant biology may seem a peaceful branch of science. Unlike toads and bees, plants do not invade places, wipe out the local wildlife, or cost us lots of money, do they? Wrong! Plants can be every bit as pesky as animals. A weed is defined as a plant growing where it is not wanted. And you cannot find much worse weeds than these.

COULD YOU COMPETE WITH KUDZU?

The Chinese vine called kudzu has been grown in Japan for centuries for its edible roots and to make paper from its stems and leaves. In 1876 this creeper was shown at an exhibition in Philadelphia, Pennsylvania. Soon Americans were planting it as a decorative "porch vine." By the 1930s, farmers grew it as food for animals and to stop river banks from being washed away.

But kudzu can grow with incredible speed—up to 30 cm in a day! Now it has spread to the American south, where it has choked huge tracts of land with its ropelike stems and leaves as big as dinner plates. Buildings and machinery lay abandoned under a jungle of throttling growth. People chop it to pieces, try to dig up its roots, spray it with chemicals, burn it with flame throwers, and even hold kudzu-clearing barbecues where the vine fuels the fire. But the weed just can't be stopped.

Get off!

PRETTY, AND PRETTY NASTY

The water hyacinth looks pretty, with lilac flowers and elegant leaves. But it's the world's worst waterweed. Helped by humans, it has spread from South America to choke thousands of rivers and lakes. In one summer, twenty-five small water hyacinths can produce two million new plants. These can cover an area of three soccer fields with a two-meter-thick tangle of leaves, roots, and stems.

This weed strangles other water plants, suffocates fish and other creatures, clogs irrigation pumps and ditches, blocks pipes and turbines in electric dams, and tangles boat propellers and dredgers. Its damage costs billions of dollars every year. One of the few ways to control this waterweed is with the help of manatees, or sea cows. They love to munch the horrible hyacinth.

I think I'm in Heaven

A Prickly Problem

The prickly pear, a type of cactus, was originally taken to Australia for use as hedges. Local animals could not stomach its spines—and soon it spread and spread. By the 1880s, over 30 million hectares of eastern Australia were a thick carpet of cacti, useless for farming. It's still a pest in some areas.

33

A Hole in the Head

In the Stone Age, people who had headaches tried not to complain too much. One supposed remedy for the ache was to chip a hole in the skull to let out evil spirits that were believed to be causing the pain. The "doctor" would strike a stone chisel with a rock to cut through the skin and bone and expose the brain. Ancient skulls with partly healed holes show that some people actually survived this treatment.

Are you sure about this?

THE HEAD-DRILL

Making holes in the head was called trephining or trepanning. In the Dark Ages, doctors did this with a metal drill. The doctor would turn the drill by pulling a bow back and forth, with its string wound round the drill. If the patients lived, they were given the largest fragments of their skull bone, and they would hang these around their neck as a lucky charm!

NOT VERY HUMOROUS

Hippocrates was the most famous doctor of Ancient Greece. He believed that good health resulted from a balance of four substances, called humors, in the body. These were blood or *sanguis*, phlegm or *pituita*, yellow bile or *chole*, and black bile or *melanchole*. Too much of one humor would change your personality and even make you ill. Too much blood, for example, made you cheerful, or sanguine. Too much black bile made you depressed, or melancholic.

PHLEGM — BLACK BILE — BLOOD — TODAY'S SPECIAL — YELLOW BILE

HIPPOCRATES
FOR ALL YOUR HUMOR NEEDS

Cheerful, or sanguine

Sluggish, or phlegmatic

Quick-tempered, or cholic

Depressed, or melancholic

Now come on chaps, it won't hurt a bit...

IT'S ONLY A FLESH WOUND

Roman doctor Claudius Galen wrote many books about the inside of the human body. Yet tradition prevented him from using a knife to cut open bodies. So how did he know what was in there? He just happened to be the official doctor for the gladiators fighting in the Coliseum of Rome. He saw plenty of innards on that job!

HELPUX!

LET THE BLOOD RUN FREE

In medieval times, people believed that many diseases were caused by impure blood, or by too much blood in the body. The obvious cure was to let some out! The doctor, therefore, would cut open a blood vessel and let the patient bleed for a while. Or he would stick wormlike leeches on the skin. These creatures then sucked out the patient's blood, because that's what leeches feed on.

Hmm... MORE leeches, I think

Did Ye Know?

In ancient India, doctors were supposed to look at the inside of the body as part of their training. Yet the law did not let them use knives on dead bodies. So instead, they would leave the dead body in water. After a week, it was so soggy that it could simply be pulled to bits!

Taking Your Medicine

Now, what seems to be the trouble? Headache and upset stomach? We won't bother to take your pulse and temperature, or measure your blood pressure. Instead, we'll consult your star chart, taste your urine, and check the weather outside. Then we'll forget it all and prescribe a potion made from toad poison and viper venom. Early medicine was certainly weird and wacky!

Mmm, nearly ready

THERIAC IS GOOD FOR YOU

Physicians spent centuries working on the recipe for theriac. This was supposed to be a cure-all, or "panacea" (named from the Greek goddess Panacea, who was the daughter of Asklepios, the god of medicine). Theriac was originally used as an antidote to snake bites, and called mithridatium, after a King Mithradates, who tested the treatment on his slaves.

The ancient Romans made theriac from more than fifty ingredients, including opium from poppies and viper meat. In the Middle Ages, physicians added even more ingredients. They said that theriac had to molder and mature for years into a thick, smelly syrup before you took it. If you lived that long . . .

EYE OF NEWT, TONGUE OF TOAD

In the Dark Ages, just about anything could be used for making medicines, potions, and ointments. If an illness affected a certain part of the body, doctors used natural ingredients that resembled the infected body part, or the disease's symptoms, to cure the problem. So liver ailments were treated by eating a plant called liverwort. Diarrhea was treated by drinking the rusty-looking water from stagnant ponds. If the patient was cured, fine. If he died, there was no one to complain!

TAKING THE URINE

From ancient times, doctors thought that a patient's urine was all they needed to make a diagnosis. The doctor would study the urine's color, clarity, runniness, frothiness, smell, and taste, and immediately diagnose the problem! The glass urine flask even became the symbol of the medical profession. In reality, this procedure, called uroscopy, has no proper scientific basis. And by the eighteenth century it was becoming less common. Today, doctors still take urine samples, but now they perform a wide range of chemical tests on them in a laboratory.

No, don't tell me, I think I know the problem

There's no place like home!

BREEDING THEN BITING

Before people understood that mosquito bites spread a disease called malaria, they thought it might be carried by small bugs who bit them at bedtime. So people put the legs of their beds in bowls of water, to drown the bugs as they tried to crawl into bed with them.

Unfortunately, mosquitoes loved these small, stagnant "ponds." They actually bred in them, right beneath the victims they would bite next! People also believed that malaria was caused by an unseen staleness or miasma in the air. Hence the name mal-aria, "bad air."

37

Alchemy

Ancient Greeks, such as Aristotle, believed that everything in the world was made of four elements—earth, air, fire, and water—combined in different proportions. Could these elements be changed, or transmuted, one into another? So began centuries of alchemy, searching for wonders like the Philosopher's Stone and the Elixir of Eternal Youth.

She doesn't look a day over 2,000

THE ELIXIR OF ETERNAL YOUTH

The search for the Elixir of Eternal Youth began in China more than 2,000 years ago. Drink it, and you would live forever and always look and feel young. The Lady of Tai kept her looks for 2,000 years. But this was because, as soon as she died in 186 B.C., she was preserved in a coffin full of brown liquid containing mercuric sulphide and methane gas.

THE PHILOSOPHER'S STONE

The main aim of alchemy was to get rich quick. What? No, sorry, it was to carry out valuable scientific research that just happened to have as its goal the transmutation of cheap, common metals like iron into valuable ones such as gold. Alchemists also searched for the Philosopher's Stone, a legendary object that would make this process easier.

EXPENSIVE FAILURE

Many kings were persuaded to pay for the alchemists' expensive research. In 1317, Pope John XXII was so fed up with alchemists, and their claims and counterclaims, he ordered that alchemy be banned. The alchemists never succeeded in their main goals. But by developing equipment and experimenting with chemicals, they contributed greatly to the study of physics, chemistry, drugs, and medicines.

Tea up!

STOP THE PRESS~Aristotle Builds Nuclear Reactor!

The basic idea of changing one substance into another has also come true—after a fashion. Modern physicists have found that unstable, radioactive substances change into stable, non-radioactive ones. For example, a form of uranium becomes thorium and then lead, and a type of cobalt turns into nickel. These types of change happen in nuclear reactors. The alchemists would have been pleased!

MERCURY MADNESS

Isaac Newton spent his later years doing experiments to find the Philosopher's Stone. Like many alchemists, he used mercury, or quicksilver, a liquid metal. It is very poisonous and can cause people to go mad. This may account for the fact that Newton behaved very oddly in his old age.

39

Truly Shocking Science

Have you ever gotten an electric shock after walking across a carpet and touching a door handle, or while getting out of a car? This is due to static electricity, which is usually caused by friction and builds up on your body. If you touch the ground or something metallic or damp, the electricity flows into it instantly. (Strong electric shocks and electricity from power supplies can kill. *Never* play with electrical equipment. See what happened to Georg Richmann!)

MAKING SHOCKS

Before 1800, scientists did not have steadily flowing electricity—what we call electric current. They made static electricity with machines called electrostatic generators, which rubbed things together, such as glass and leather. The charge was stored in a device called a Leyden jar. If you touched it—*Bang!*—you could be knocked to the floor by the shock.

MEASURING SHOCKS

During the 1700s there were few machines or instruments for measuring the strength of an electric charge. So some scientists used the human body as a gauge. They measured how far the electric shock went up their arm, through their body, or along a row of people holding hands! In the 1730s, Stephen Gray, an experimenter, measured how electricity flowed through the human body. Not surprisingly, he had a lot of trouble getting volunteers.

SHOCKS CAN KILL

By the 1750s, scientists discovered that a lightning bolt is a gigantic burst of electricity. In America, in about 1752, Benjamin Franklin flew a kite in a thundercloud. The electric charge came down the damp string, through a key tied to it, and into a Leyden jar. Franklin was incredibly lucky. In Russia, Georg Richmann copied the experiment and was killed by a huge shock.

Remember NEVER try this at home, folks!

SHOCKS FOR KICKS

By the mid-eighteenth century, machines that worked by static electricity had become popular for entertainment. On stage, showmen made huge sparks to amaze their audiences, and they shocked dozens of people at the same time. In homes, the first electrical games caused people to jump and made their hair stand on end.

MORE SHOCKS

Electric current was understood and contained in 1800, when Italian Alessandro Volta invented the electrical cell, or battery. Volta disagreed with Luigi Galvani about the nature of electricity. In Galvani's experiments, dissected frogs' legs twitched when touched with a metal knife. Galvani said electricity was made inside the animals. Volta thought that electricity was made by combining chemicals, and that the animal's muscles twitched when electricity flowed through them. In a sense, both were right.

41

Suffering for Science

When scientists invent new machines or processes, someone has to try them out. Many times scientists do the testing themselves. They become "human guinea pigs" and take the initial risks. But sometimes other people get to be the brave pioneers—whether they like it or not!

You're always hungry

THERE'S A HOLE IN MY STOMACH

In 1822, a Canadian adventurer named Alexis St. Martin was accidentally shot in the stomach. His terrible wound was treated by U.S. Army doctor William Beaumont. It healed—but with a hole from the outside right into the stomach! For the next seven years, St. Martin allowed Beaumont to push tubes and pads through the hole into his stomach so the doctor could collect and study digestive juices and their effects on food.

Beaumont published his results in a book called *Observations on the Gastric Juice and the Physiology of Digestion*, which helped medical research enormously. He lived to sixty-eight years of age. St. Martin, still with his stomach hole, reached eighty-two!

Hooray!

BAILING OUT

In 1849, a worried ten-year-old boy sat in a strange contraption with three sets of wings and a boatlike body. This was a glider, the invention of George Cayley, a pioneer of aircraft design. The craft was towed on a string, rose like a wobbly kite into the air, and came down with a bump. The boy declined to have another ride.

Four years later, when he was eighty, Cayley made a much larger glider. He instructed his coachman to pilot it. The glider, which had no controls, flew unsteadily before crash-landing. The shaken coachman quit on the spot, saying: "I was hired to drive, not to fly."

The First Vaccine

Smallpox is a terrible disease that has ravaged the world for centuries, killing millions of people. Cows get a similar but milder disease, called cowpox. It was believed that cowherds who caught cowpox were then protected against the far more serious smallpox.

In 1796, English physician Edward Jenner decided to test this belief. He took some fluid from a cowpox sore on the hand of a dairy maid named Sarah Nelmes and pricked it beneath the skin of a healthy boy, James Phipps. Six weeks later he deliberately gave James smallpox. If the boy had died, Jenner would have been called a criminal, not a hero. But Phipps did not develop the disease, and Jenner became famous. This treatment was accepted as the scientific beginning of vaccination, or immunization, to protect against disease.

Saved from Rabies

Rabies is a fatal disease, caused by a virus and spread by animal bites. In 1885, French medical scientist Louis Pasteur was researching a vaccine against it. One day after a young boy named Joseph Meister was bitten fourteen times by a mad dog with rabies, Pasteur was persuaded to try the vaccine for the first time. Joseph received twelve painful injections —but he lived. Soon thousands of people were being saved from a painful death.

When Joseph was older, he worked as a guard at the Pasteur Institute in Paris.

During World War II, he would not let enemy soldiers into the Institute—and he was killed for his loyalty.

Beyond the Call of Duty

The first person ever to steer a glider in flight was German engineer Otto Lilienthal, in 1891. In five years he made over 2,500 glider flights, before a crash in 1896 finally claimed his life. His gravestone is engraved with his dying words: Sacrifices Have to Be Made. Here are some more dedicated scientists who suffered in their search for knowledge and progress.

isdooo

Frankly, you should lose some weight

A Life in the Ballance

Italian professor Santorio Santorio, a colleague of Galileo, was obsessed by the workings of the human body. To study it, from about 1590 to 1620, he spent much of his time in a *Ballance*, a tiny room suspended from giant scales. He ate, slept, exercised, washed, and read there. He weighed all the food and drink going into him, and weighed and studied all the products coming out, including exhaled moisture, sweat, urine, and feces!

Can You Stomach It?

In the mid-1700s, Italian scientist Lazzaro Spallanzani carried out many important experiments on digestion. In one set of tests, he swallowed sponges on strings, pulling them back out after they had absorbed his stomach juices. Then he added different foods to the sponges and kept them warm under his armpits to see how digestion happened. He even went to church services with these samples!

Sir, dinner is served

Breakfast

CRITICAL CHEMISTRY

Versatile Frenchman Antoine Lavoisier founded the modern science of chemistry. He showed that air contained two main gases, nitrogen and oxygen, and he devised the symbols we use for chemicals, such as H_2O for water. He also improved French farming and town lighting, education, mapmaking, and the tax and banking systems.

In 1789 the French Revolution began. Among the leaders was Jean Paul Marat, who had been a doctor and scientist. Unfortunately, Lavoisier had once criticized one of Marat's publications! The revolutionaries also did not like Lavoisier's involvement in taxes and banks. So Lavoisier was put on trial and guillotined the next day.

I'll finish it later

CUT YOUR TAXES

LA CHIMIE par J.P. Marat

BUNSEN'S BURNING

German chemist Robert Bunsen carried out hundreds of experiments and invented dozens of laboratory gadgets, such as the spectroscope and filter pump. But he suffered greatly during his work.

In the 1840s, he decided to make some compounds called cacodyls, which contained arsenic. Bad choice! They are poisonous, smell awful, and catch fire easily. First, Bunsen lost an eye in an explosion. Then he got arsenic poisoning, which gave him muscle cramps, severe diarrhea, and partial paralysis. As a result, he moved on to safer work—taking gas samples from volcanoes and inside blast furnaces!

The Bunsen Burner

Oddly, the gas burner named after Bunsen was not his invention. It was probably developed by his assistant, Peter Desdega, from an even earlier design.

Also, Bunsen cared little for his appearance or personal hygiene. One lady said Bunsen was charming, and she would like to kiss him— but she would have to wash him first.

Flapping and Falling

To fly like a bird—that has been the dream of people through the ages. Even now, we cannot flap through the air like birds, and so we have built many machines to help us fly, such as balloons, airships, gliders, and powered airplanes. But the fight for flight has been a long and difficult journey, with many broken limbs and expensive failures along the way.

WRONG FEATHERS

In the eleventh century, a monk named Eilmer tried to glide from a tower, on wings made of bird feathers. He crashed and broke both of his legs. Eilmer blamed the wings, which were made partly of chicken feathers, because chickens are poor fliers!

Err, aukk, squawk

IN A SPIN

In the early 1500s, all-around genius Leonardo da Vinci sketched designs for human-powered flying machines. One of them was an early type of helicopter, with rotating screwlike wings. But it did not have a stabilizing rotor at the rear, like a modern helicopter. Without this stabilizer, the wings would have spun one way, and the pilot would have spun almost as fast in the opposite direction!

Err, ooh

What a flapping cheat

In 1809, Joseph Degen claimed he was the first man to fly like a bird. When he flapped wings attached to his arms, he appeared to fly through the air with the greatest of ease. But he cheated—he was hanging by a rope from a large balloon!

IS IT A BIRD? IS IT A PLANE?

No, it's a big flop. In 1903, famous American scientist Professor Samuel Langley spent $70,000 on a four-winged airplane, the *Great Aerodrome*, which looked like a giant dragonfly. In October, it was catapulted along take-off rails mounted on a houseboat in the Potomac River in Washington, D.C. But instead of flying, the craft flopped straight into the water, and pilot Charles Manley almost drowned.

Langley and Manley soon made a fresh attempt—and another immediate splash-crash-landing. The newspapers gleefully poked fun at their very expensive failure. But not long after . . .

. . . FLIGHT AT LAST

On December 17, 1903, the first true self-powered airplane flew, and it cost about $40. The *Flyer No.1* was built by bicycling brothers Wilbur and Orville Wright. The Wrights made four successful flights on the windswept coastal sands of Kitty Hawk, North Carolina. The brothers planned more flights, but strong winds tipped the *Flyer,* and it was bumped and bounced along the sands. The first successful plane was bent and broken only minutes after its great triumph.

Err, well done, Orville!

Chocks Away!

Since the Wright Brothers made the first powered flight, there have been many strange ideas about airplane design. No matter how many calculations they did, or how many models they tested, designers never quite knew what would happen until the first test flight. Would *you* have gone for a trip in one of *these* weird-winged wonders?

Too Many Wings

If one wing moving through the air could provide enough force to lift a plane, then surely lots of wings would be far better? The Multiplane took this concept to extremes with twenty thin wings, like a flying Venetian blind. And fly it did, in 1907, but only for three short hops. On the fourth attempt, piloted by its designer, Horatio Phillips, it came down with a bump and crumpled into a heap.

I have a dream...

Too Fast

The Messerschmitt Me 163 was a World War II fighter with a difference. It blasted at great speed high into the air, using a rocket engine powered by chemicals. However, the take-off used up all its fuel, and the pilot then had to glide the plane back to the ground. The pilot was lucky if he or she got one high-speed swoop past an enemy plane. And worse, if the plane's caustic fuel leaked from its tank, it would injure or kill the pilot in seconds.

Tally-ho! Wunderbar!

Too Heavy

The enormous Hercules H4 flying boat was designed by American millionaire Howard Hughes. It was nicknamed the "Spruce Goose" because its frame was made of spruce wood. It flew only once, in 1947. Even eight engines could not lift the vast body properly. Now in a museum, it holds the record for the longest wingspan of any plane, at 97.5 meters. In contrast, the wingspan of the Wrights' *Flyer* was just 12 meters.

Too Slow

In 1954, the Convair company claimed that their F-102 Delta Dagger would be the first missile-carrying jet fighter to fly at supersonic speed—faster than sound—in level flight. Tests on models in wind tunnels proved it could be done. But when the first F-102 took to the sky, it could not break the sound barrier. Embarrassed designers found that the shape of the plane was not smooth enough. They admitted that models sometimes fly better than the real thing and had to redesign it with a better shape.

Too Dangerous

Lockheed's F-104 Starfighter was a super-streamlined jet fighter that first flew in 1954. Pilots appreciated the need for its downward-firing ejector seat when flying high—it would throw them clear of the Starfighter's very high tail. But what would happen if they had to eject during take-off or landing? Eventually a more powerful upward-firing seat was fitted. But in the 1960s, over 120 Starfighters crashed on routine flights, and many pilots were killed. Worried airmen called this plane the "Flying Coffin."

Pheew! That was a really close shave!

Mis-Construction

Large construction and engineering projects involve lots of experts—including planners, scientists, designers, and engineers. They consider vital questions. Is the project being conducted in the right location? Will it work properly and reliably? And can it survive storms, earth tremors, and other catastrophic events? But despite everyone's best efforts, things still go wrong.

THE TAY DISASTER

The Tay bridge near Dundee, Scotland, was the world's longest bridge. However, the designer had fatally under-estimated the force of wind blowing against it. In a howling winter gale in 1879, the bridge's central section blew down as a train was crossing. The train and seventy-five passengers crashed into the dark water below.

GALLOPING GERTIE SHAKES HER STUFF

The Tacoma Narrows bridge over Puget Sound, in Washington State, opened in July 1940. This suspension bridge, with a main span of 853 meters, was beautifully slim and elegant. Unlike the Tay bridge, it was designed to flex slightly with the winds. Local people nicknamed it "Galloping Gertie" since it would sway in even a light breeze.

On November 7, a medium wind caused the road deck to begin to ripple and twist. Soon the undulations got bigger. People driving across in their cars got out and ran. In a few minutes the road was bending as if made of rubber. Finally, the bridge tore itself to pieces and crashed into the river. One cause of the collapse was the design of the road deck. As it twisted in the wind, its shape created increasingly large up and down movements.

I got rhythm

THAT SINKING FEELING

In the early 1990s, Japanese engineers built the new Kansai airport on an artificial island. Unfortunately, the island began to sink into the soft seabed, and the runway is now just five meters above the water. Engineers believe that the island will eventually stabilize, but they have had to use 900 hydraulic jacks to support the airport terminal!

MORE BUILDING BLUNDERS

2580 B.C. Halfway up, builders had to reduce the angle of Pharaoh Sneferu's new pyramid. Today, it is known as the Bent Pyramid.

1968 New trains in Long Island, New York, were found to be too big to pass safely through the tunnels.

1970 A calculation error meant one pre-made section of the Westgate bridge in Melbourne, Australia, was built too short. When engineers tried to plug the gap with concrete blocks, the bridge collapsed, killing thirty-five workers.

SHOCKINGLY STRONG

New York's Empire State Building was the tallest in the world when it was completed in 1931. On a foggy day in 1945, a B-25 Mitchell bomber flew straight into the side of the giant skyscraper. One of the plane's engines smashed right through the building, and many people were killed or injured.

But this was also one case where the builders had gotten it right. The structure was so strong that the massive steel inner skeleton was almost undamaged. Today, engineers say that the Empire State Building could have a steel frame less than half the weight, and still be safe.

False Starts

Imagine life without the car. Modern road vehicles are central to daily life. We take for granted the efforts of engineers and designers as they improve our sophisticated cars—striving for better engines, aerodynamic shapes, and more comfortable interiors. However, the story of the automobile has also had its fair share of surprises, shocks, and false starts.

GETTING UP STEAM

The first self-powered vehicle—called an automobile—was Nicolas-Joseph Cugnot's 1769 "cannon tractor," designed to pull gun carriages or carry four people. But this steam-engined three-wheeler went slower than walking pace and had to stop every minute to pick up steam. During a French army demonstration in 1771, it careened out of control, crashed into a wall, and turned upside down. The generals decided to stick with horses.

RIDING ON AIR

The first gas-powered car with pneumatic, or air-filled, tires was a Peugeot. In 1895 the Michelin brothers entered it in the Paris-Bordeaux-Paris race. They used 22 spare inner tubes and mended at least 100 punctures before finally giving up after 90 hours. The race had been won 41 hours before, by a car with standard solid rubber tires.

WORLD SPEED SHOCKS

The vehicles that held the first six world land speed records were not powered by gas engines, like modern cars, but by electric batteries. From 1898 to 1899 they increased the record from 39 miles/hr to 66 miles/hr. As well as the danger of speeds previously unthought of, the drivers faced another risk—several of them received electric shocks from their machines!

Err...

BACK FROM THE DRAWING BOARD

Today, record-breaking hopefuls spend fortunes and many years developing the best technology that money can buy. Yet in 1924, Ernest Eldridge took an ordinary Fiat car, souped it up a bit, and reached 146 miles/hr—carrying the extra weight of a passenger, on worn tires, and on an ordinary tree-lined road near Paris! But his speed didn't count because his car didn't have a reverse gear, which broke speed record rules. So he added a reverse gear and took the record at 144 miles/hr!

Left on the right

Why do Europeans drive on the right side of the road? Because Napoleon Bonaparte told them to, in 1807. His idea was to prevent accidents when two horse-drawn carriages thundered towards each other. Why the right? Perhaps because the British—his sworn enemies—had always used the left side. Vive la différence!

A Life on the Ocean - Glug!

It's a marvelous life on the ocean wave. As you relax, far from school or work, the sun shines from a clear blue sky onto a calm green sea, and the boat rocks gently. Dolphins leap and play, and the long-winged albatross soars above. Wonderful! Unless, that is, you're on one of these unusual or unfortunate vessels.

BENT AND TWISTED

In 1863, the *Connector* had a clever design. Its three sections were joined by two sets of huge hinges. The two cargo sections could be detached for loading and unloading (like train cars), while the rearmost, powered section went off to do other jobs. The hinges also allowed this long ship to bend up and down in high waves. Tests on a smooth, calm river went well, but they showed the design would not work in a rough sea. The ship would twist in high waves and snap the hinges. The *Connector* was never used.

ROUND AND ROUND

In the 1870s, the Russian navy built two circular warships, the *Admiral Popov* and *Novgorod*. They were designed to swing around quickly so that their huge guns could point anywhere, and to stay steady so the aim was accurate. When the guns fired, their recoil would not tip the ship to one side. This part of the design worked well. But the ships were slow, and even small waves splashed over them, flooding the decks and damaging their plating. And in a strong current, they spun round like tops!

I said forward, man ~FORWARD!

ROCKED AND ROLLED

When the wind blows and the waves swell, a ship rocks and lurches. The unpleasant result: seasickness. The *Bessemer*, built in 1875, was designed by steelmaker Henry Bessemer, who suffered greatly from seasickness. The *Bessemer* had a passenger portion on guide rails that could tilt inside the outer hull. As the hull rocked and rolled, a "steersman" controlled hydraulic machinery to keep the passenger part level. It worked fine in small, predictable waves. But the steersman could not predict big, irregular waves, and the anti-roll machinery was simply too slow. As a result, passengers felt even more seasick than usual.

The *Bessemer* also had two front ends, so it did not have to turn around in port. But it crashed into a wall at Calais harbor anyway, and was wrecked.

You always did drive too fast!

THE UNSINKABLE TITANIC

In April 1912, the *Titanic* set sail on its first voyage from Southampton, England, to New York. It was the world's largest and most luxurious passenger liner. It was also designed to be unsinkable: it had a very thick, stiff hull and watertight compartments inside. "Even God couldn't sink her," said one of the ship's owners. But, near Newfoundland, the liner hit an iceberg, which ripped open several of its compartments—and the *Titanic* sank.

Of the 2,200 crew and passengers, about 1,500 died. The sinking of the *Titanic* was one of the worst sea disasters of all time.

The loss of life could have been avoided. The *Titanic* was going too fast in an area known to contain icebergs. And, because the great liner was thought to be unsinkable, it did not carry enough lifeboats for everyone on board.

Inventions

Every year, hundreds of hopeful inventors and scientists unleash their discoveries on the public. Very rarely, one of these inventions is cheap to make, easy to use, reliable, fills a real need—and so becomes a part of daily life. Here are some inventions that took a while to be perfected, as well as a couple of unusual ones that were devised by famous scientists.

ALL SCREWED UP

The basic screw shape was supposedly invented by Archimedes, an ancient Greek mathematician. Later, in the sixteenth century, someone actually made a screw for woodworking—a headless nail with a twist along its length. Unfortunately, no one had yet invented the screwdriver. So you hammered it in, and there it stayed. Neither the screw with a slotted head, nor the screwdriver with which to twist it in or take it out, was invented until one hundred years later.

CAN THE CAN BE CAN-OPENED?

Some claim that the metal "tin can" was invented in France in 1804 by Nicolas Appert, to store food for Napoleon's soldiers. Others give credit to Englishmen Donkin and Hall in 1811. Either way, the can was stronger than the glass preserving jars used previously.

Unfortunately, the lever-jawed can opener was not invented until 1855, by Robert Yeates. Before this, people had to use a hammer and chisel.

SHOUT LOUDER, EDDIE—I'M SEWING

American Thomas Edison's inventions changed daily life—the light bulb, phonograph, electricity power station and distribution grid, improved telegraph, and better telephone. But his sound-powered sewing machine of the 1880s was less successful. To make it work, you had to shout nonstop at the top of your voice, into a mouthpiece. It was much easier to use the normal foot pedals. Quieter, too!

Isaac's Pussy's Push-Up PORTAL

Very few scientists deserve the label "genius." One is Sir Isaac Newton, who laid the foundations of modern physics with his laws of gravity and motion. His *Principia* (1686) is one of the greatest science books ever written. Isaac also invented the cat door!

SEW!

WHICH BRIGHT SPARK INVENTED THIS?

For centuries, people used heavy irons to uncrease and smooth their clothes. The iron was heated near a fire or in hot water. When electricity began to arrive in houses, in about 1890, all manner of new electrical gadgets appeared. The first electric iron was heated by an arc, an intense spark jumping continuously between two carbon rods inside the iron. It snapped and crackled fiercely. It could also brighten up the day!

Perhaps, however, it was an improvement on an iron that was heated by burning gasoline . . .

Just an Accident

Scientists make mistakes and have accidents, like anyone else. Usually this means they have to begin anew. But occasionally an accident has led to a lucky break and a great new discovery.

PERKIN'S PURPLE PATCH

In 1856, a clever chemistry student named William Perkin was working in his home laboratory. His attempt to make the drug quinine from a substance called aniline failed horribly. He was left with a disgusting black goo. So he tried boiling it—and it turned a brilliant purple! Perkin realized this substance, later called mauve, could be used to color, or dye things. Perkin's mauve was the beginning of a huge new industry, making dyes from chemicals in the laboratory. By the age of thirty-six he was able to retire, a rich man.

X FOR UNKNOWN

In 1895, German physicist Wilhelm Roentgen was experimenting with a discharge tube (a forerunner of the television set). Roentgen noticed that a piece of a chemical-coated card near the tube glowed when the tube was switched on.

Tests showed that a new type of invisible ray was coming from the tube. It could travel through wood, thin metal, and human flesh, but it was stopped by lead or bone. The new rays could be detected by photographic paper, so Roentgen used them to take a photograph of the bones in his wife's hand. For a time they were called Roentgen's rays, but today we know them by the name that Roentgen gave them— X-rays, the "X" for "unknown."

I thought it was quite a good likeness

A Strange Activity

In 1896, French scientist Antoine Becquerel noticed that some of his photographic paper had become dark while still in its light-proof wrapping. Puzzled, Becquerel realized that the paper had been stored near a chemical containing uranium. Tests showed the uranium was giving off mysterious rays. Perhaps they were the new rays discovered the previous year by Roentgen? Actually, no—Becquerel had discovered yet another type of ray.

We now know these rays as "radioactivity." This name was invented by Becquerel's colleague, Marie Curie, who later discovered the radioactive substances polonium and radium.

How very interesting!

Killer Mold Kills Killer Germs

In 1928, Scottish scientist Alexander Fleming accidentally left the lid off a small round dish of germs he was studying. Before long he noticed that the germs on the dish were dying. Fleming found that some microscopic spores of a fungus, or mold, had floated into the dish—and were killing the bacteria! Fleming identified the mold as *Penicillium*, and he called the germ-killing chemical it made "penicillin."

Fleming's work was taken up by Australian medical scientist Howard Florey and German chemist Ernst Chain. They discovered how to make large amounts of penicillin. It became the first antibiotic, or bacteria-killing drug, and it was used during World War II to treat the infected wounds of soldiers. Since then, penicillin and other antibiotics have been used to save millions of lives.

Mad, Bad, and Sad

Good science does not necessarily mean good sense. Some great scientists have been more than a little crazy. They thought clearly and sensibly about their work, but they were far less successful in other areas of life. Perhaps a dash of madness, badness, or sadness is the price of a stroke of genius.

THE MAD MINERALOGIST

Rock-and-mineral expert William Buckland was one of the first people to reconstruct—rebuild—prehistoric animals from their fossilized remains. He wrote down the very first description of a dinosaur, which he called *Megalosaurus*. In 1824 Buckland suggested that the animal was a giant meat-eating lizard—an inspired guess, since the word "dinosaur" did not enter the common vernacular until 1841.

Buckland was strangely keen on coprolites, rock-hard droppings from prehistoric animals. He liked to shock friends by stroking the fossil feces, and his dining table was even made from huge rocks of dino-dung. Buckland also wanted to try eating one of every creature in the whole animal kingdom! He said mole meat had the worst taste, followed by bluebottle flies. He tasted mice dipped in batter, as well as crocodile, jackal, and many other unusual dishes.

The Stomach, Sir, rules the World!

excuse me...

...suddenly I don't feel hungry

...so take two *#~## swords twice a *#~## day!

THE BAD-LANGUAGE DOCTOR

Paracelsus was a famous sixteenth-century doctor in Basel, Switzerland. (His real name was Philippus Aureolus Theophrastus Bombastus von Hohenheim, but let's stick to the shorter one he liked to use.) He spoke out against the horrific treatments of the day, such as bleeding patients almost to death and giving them dangerous drug mixtures of poisonous plants and animal venoms. He wanted to use medicines more scientifically. He tried them one at a time, in a logical fashion, to test if they worked. This was the first serious medical drug research.

But Paracelsus also had a strange side. He said that the pain of a toothache could be moved into a tree, and that a sword wound was cured by rubbing ointment on the sword. He also believed in gnomes and nymphs. And Paracelsus swore so much, and so offensively, that he was forced to leave Basel forever.

ooh...err...

THE SADLY RECLUSIVE CHEMIST

Henry Cavendish made some very important discoveries in chemistry. He found that water was made of two substances, hydrogen and oxygen, and that air contains many gases. Cavendish also invented ways of weighing gases, studied heat and electricity, and measured the force of gravity and the density of Earth.

But Cavendish was eccentric and painfully shy. He avoided strangers and could hardly stay in the same room with a woman. He was very rich, but he dressed in shabby clothes and ate stale food, alone in his room. He published little of his scholarly work, and in 1810 he sadly died alone.

61

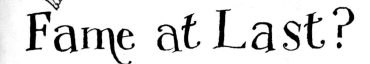

Fame at Last?

The history of science is full of great discoveries and inventions. But the achievements of many brilliant scientists were not appreciated during the scientists' lifetimes. Some never found fame—or fortune. Others got into trouble because of their work. Today, with the benefit of hindsight, we can recognize their accomplishments.

Nice try!

THE THIRD BALLOONIST

On November 21, 1783, Frenchmen Pilâtre de Rozier and François Laurent became the first fliers, in the Montgolfier brothers' hot-air balloon. This was a blow to physicist Jacques Charles. Three months earlier he had tested a small balloon filled with lighter-than-air hydrogen gas. It had flown 25 km—but, upon landing, villagers had torn it to pieces, thinking it was a monster! Eventually, on December 1, Charles made and flew in a larger balloon. His name lives on, though, since he devised Charles' Law, linking a gas's temperature, pressure, and volume.

There... that's much better

TO UPSET A KING

Long before Jean (Luc) Picard, captain of the USS *Enterprise* in *Star Trek: The Next Generation*, another Jean Picard was traveling, around France. He wanted to calculate the earth's size by measuring how much sections of land curved, using instruments he had invented himself. As we now know, his results were very accurate. Picard then drew the first detailed map of France, including all the coastlines. In 1682 he proudly showed it to his king, Louis XIV. Louis compared it to the previous version and got quite angry. His kingdom was much smaller than he would have liked to believe!

It's mine!

TV, or Not TV?
Scotsman John Logie Baird invented the television. True? Er, well . . . In the 1920s he invented a system for transmitting moving pictures. But his system, which used a quickly-rotating scanner, did not work well enough. The all-electronic TV that we use today had its beginnings in the iconoscope, developed in 1924 by Russian-American scientist Vladimir Zworykin.

MORE PEAS, BROTHER?
Gregor Mendel was an Austrian monk who grew peas in his monastery garden. He bred different shapes and colors of pea plants by transferring the pollen from one flower to another. His painstaking work led to a great scientific discovery—the laws of genetics and inheritance, which he described in 1856. These laws apply to all plants and animals, including humans. They explain why features such as hair color and eye color run in families.

Mendel proudly showed his results to experts such as leading German botanist Carl Naegli. But they were unimpressed and advised him to breed more pea plants. "Pardon?" said Gregor, "I have already studied 21,000!" Sadly, he was sent away, and his research was ignored. Only in 1900, sixteen years after his death, did biologists recognize the tremendous importance of his work. Like many scientists before and since, poor Gregor had learned that the search for knowledge has its ups, downs, and ins and outs—In the world of *Shocking Science*.

Bill

Ben

You don't look much like your father

oh no, peas for lunch again

INDEX

A
Abominable Snowman, 15
airplanes, 21, 47, 48–49, 51
alchemy, 38–39
Alhazen, 16
aliens, 19, 20–21
antibiotics, 59
Archimedes, 56
Aristotle, 8, 38
astronomy, 4, 8–9, 18–19, 23

B
Baird, John Logie, 63
balloons, 46, 62
battery, 41
Becquerel, Antoine, 59
Bible, 6, 7, 24
Big Bang, 7
Bigfoot, 15
bloodletting, 35
Bonaparte, Napoleon, 53, 56
bridges, 50
Buckland, William, 60
Buffon, Comte de, 7
Bunsen, Robert, 45

C
can opener, 56
cars, early, 52–53
cat flap, 57
Catastrophe Theory, 7
Cavendish, Henry, 61
Cayley, George, 42
Charles, Jacques, 62
chemistry, 38–39, 45, 61
cold fusion, 5
Columbus, Christopher, 11, 15
Continental Drift, 17
Cope, Edward Drinker, 26–27
Copernicus, Nicolaus, 9
corn circles, 21
Cosmas, 13
cowpox, 43
cows, 20, 43
Curie, Marie, 59
Cuvier, Baron Georges, 24

D
digestion, 42, 44
dinosaurs, 26–27, 29, 60
doctors, 35, 36, 61
dog, 5, 22, 31, 43; -headed people, 14
dyes, synthetic, 58

E
Earth, 8–9, 17, 24, 61; age of, 6–7; flat, 10–11, 12
Edison, Thomas, 57
electricity, 40–41, 57, 61
Elixir of Eternal Youth, 38
engineering disasters, 50–51
Eratosthenes, 10

F
Fleming, Alexander, 59
flight, 42, 44, 46–47, 48–49, 51, 62
fossils, 24–25, 26–27, 60; fake, 28–29
Franklin, Benjamin, 41

G
Gagarin, Yuri, 22
Galen, Claudius, 35
Galilei, Galileo, 9, 44
Galvani, Luigi, 41
genetics, 4, 63
gliders, 42, 44, 46
gravity, 10, 57, 61

H
Hippocrates, 34
hoax, 5, 21, 28, 29
humors, 34
Hutton, James, 7

I
immunization, 43
inheritance, 63
inventions, 56–57

J
Jenner, Edward, 43

K
K2, 17
Kepler, Johannes, 9
killer bees, 31
kudzu, 32

L
Langley, Samuel, 47
Lavoisier, Antoine, 45
Leyden jar, 40
Lightfoot, John, 6
Lilienthal, Otto, 44
Loch Ness Monster, 29
Lowell, Percival, 18

M
madness, 16, 39, 43, 60
malaria, 37
Mantell, Gideon, 26
Marat, Jean Paul, 45
maps, 10–11, 12–13, 16–17, 45, 62
Marconi, Guglielmo, 19
Mars, 18–19
Marsh, Othniel Charles, 26–27
Martians, 18–19
medicine, 4, 5, 34–35, 36–37, 39, 42, 43, 59, 61
Mendel, Gregor, 63
mermaids, 13
Michelin Brothers, 52
monsters, 14–15, 29, 62
Montgolfier brothers, 62
Mount Everest, 17

N
Newton, Sir Isaac, 5, 39, 57
nuclear power, 5, 39

P
Paracelsus, 61

Pasteur, Louis, 43
penicillin, 59
Perkin, William, 58
Philosopher's Stone, 38, 39
Piltdown Man, 28
planets, 6, 7, 8–9, 12, 18–19, 20–21
plants, 4, 32–33, 36, 61, 63
Pliny the Elder, 14
Polo, Marco, 14
Ptolemy, 8, 9, 10
pyramids, 18, 51

R
rabies, 43
radio, 19
radioactivity, 59
rock formation, 7
Roentgen, Wilhelm, 58, 59

S
Santorio, Santorio, 44
Sasquatch, 15
satellites, 22
Schiaparelli, Giovanni, 18
screws, 56. *See also* wrench
seasickness prevention, 55
sewing machine, 57
ships, 13, 54–55
Sirens, 13
smallpox, 43
space probes, 4, 18–19
space travel, 20–21, 22–23
Spallanzani, Lazzaro, 44
speed records, 53
Spruce Goose, 49
steam-powered automobile, 52

T
telescope, 9, 18; Hubble space, 23
television, 58, 63
theriac, 36
Thomson, William, 7
tin can, 56
Titanic, 55
toads, 36; cane, 30–31
transplants, 5
tires, 52

U
UFOs, 20–21
uroscopy, 37
Ussher, James, 6

V
vaccination, 43
Vinci, Leonardo da, 46
Volta, Alessandro, 41

W
water hyacinth, 33
Wegener, Alfred, 17
wrench, 22
Wright Brothers, 47, 48

XYZ
X-rays, 58
Yeates, Robert, 56
Yeti. *See* Abominable Snowman
Zworykin, Vladimir, 63

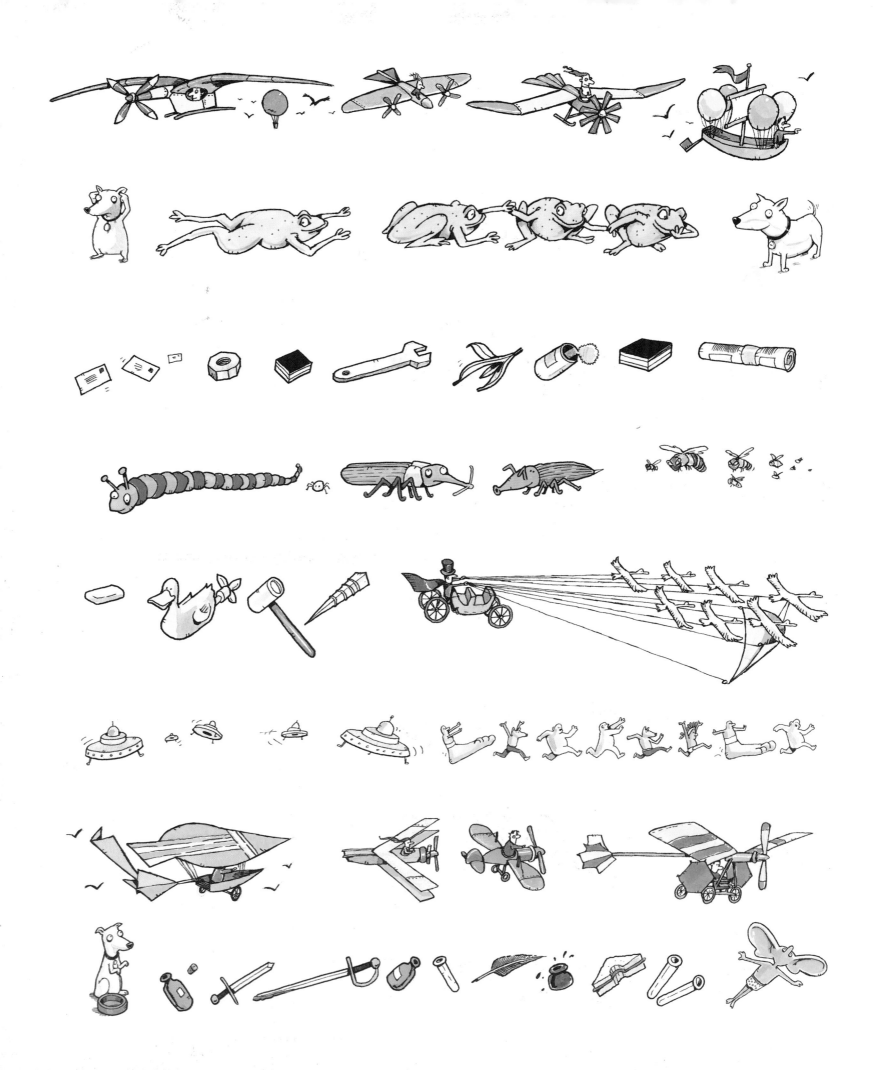